STARTING OUT

baby SKUNKS

MEG GREVE

CREATIVE EDUCATION • CREATIVE PAPERBACKS

CONT

ENTS

I AM A KIT.

I am a baby skunk.

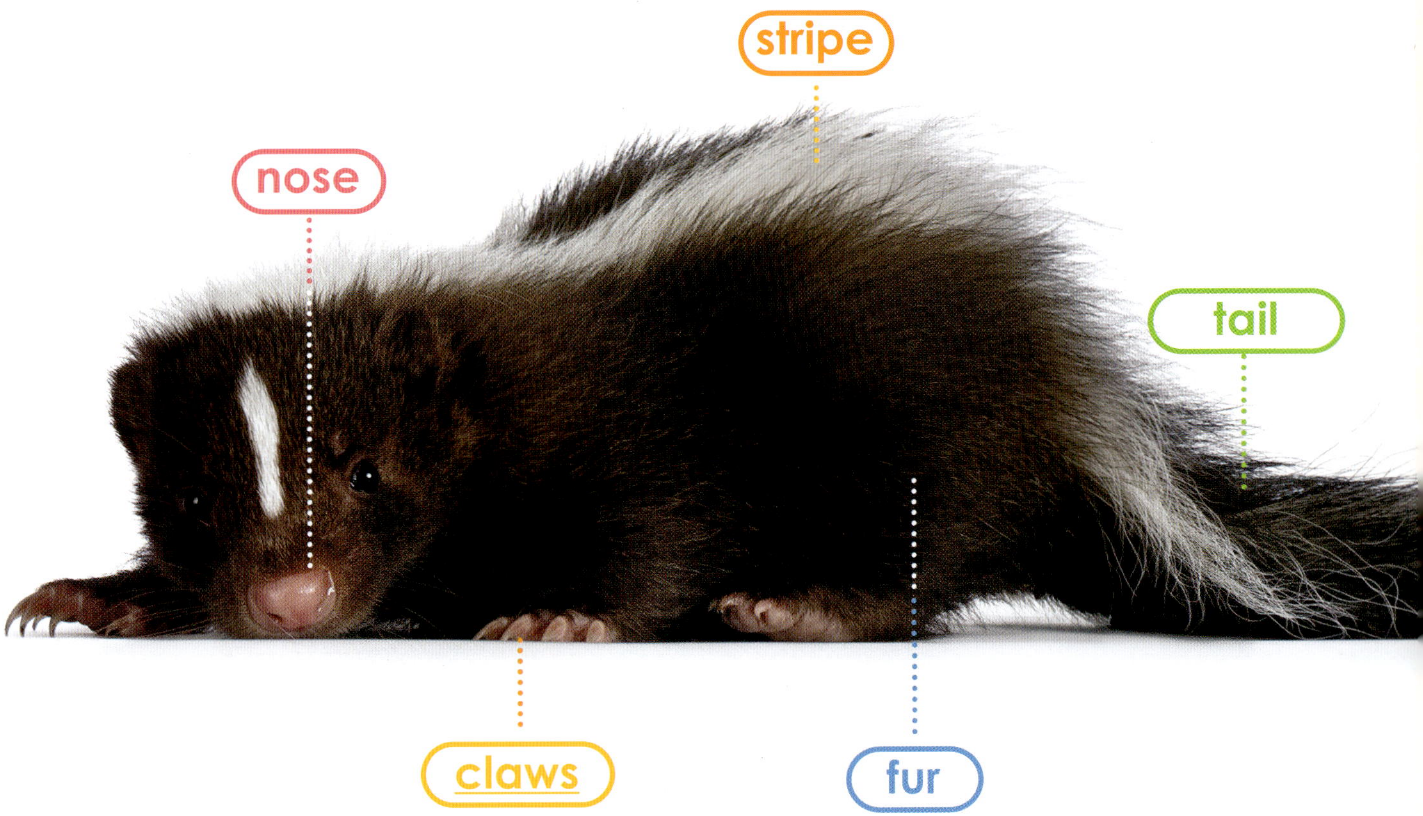

See my stripes? They tell other animals to stay away!

I was born in a warm den
lined with leaves.

My mom picks me up.
I drink her milk. I grow.

In five weeks, I leave the den.

I am ready to explore.

My nose
is good at
smelling lots
of things!

I learn to
find food.

I eat plants, insects, and eggs.
When I am scared, I lift my tail. I let out a spray. Stinky!

Now I can protect myself.

SPEAK AND LISTEN

EE-EE

EE-EE!

Can you speak like a kit?

Baby skunks squeak and chirp.

Listen to these sounds:

https://www.youtube.com/
watch?v=1ZSOQF7iylY

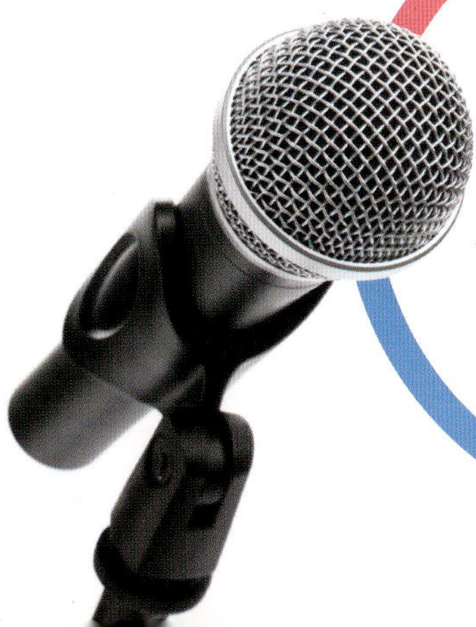

Now it is your turn!

KIT WORDS

claws: hard, sharp nails on a skunk's feet

den: a home for kits in a safe place such as under rocks or in a hollow tree

protect: keep from harm; keep safe

spray: a smelly oil that skunks shoot out from glands under their tails

READING CORNER

Bodden, Valerie. *Skunks (Amazing Animals)*. Mankato, Minn.: Creative Paperbacks, 2023.

Chanez, Katie. *Skunk Kits in the Wild*. Minneapolis: Jump! Inc., 2023.

Podmorow, Ava. *Skunks: Animals in the City*. Oliver, Canada: Engage Books, 2022.

INDEX

PUBLISHED BY CREATIVE EDUCATION AND CREATIVE PAPERBACKS
P.O. Box 227, Mankato, Minnesota 56002
Creative Education and Creative Paperbacks are imprints of The Creative Company
www.thecreativecompany.us

LIBRARY OF CONGRESS CATALOGING-IN-PUBLICATION DATA
Names: Greve, Meg, author.
Title: Baby skunks / Meg Greve.
Description: Mankato, Minnesota : Creative Education and Creative Paperbacks, [2025] | Series: Starting out | Includes bibliographical references and index. | Audience: Ages 4-7 | Audience: Grades K-1 | Summary: "Introduce beginning readers to the pungent world of baby skunks with this life science starter. Includes photos, a labeled animal diagram, "Make a Noise" section, glossary, and further resources"-- Provided by publisher.
Identifiers: LCCN 2024014700 (print) | LCCN 2024014701 (ebook) | ISBN 9798889892106 (library binding) | ISBN 9781682775752 (paperback) | ISBN 9798889893202 (ebook)
Subjects: LCSH: Skunks--Juvenile literature. | Skunks--Infancy--Juvenile literature.
Classification: LCC QL737.C248 G748 2025 (print) | LCC QL737.C248 (ebook) | DDC 599.76/81392--dc23/eng/20240419
LC record available at https://lccn.loc.gov/2024014700
LC ebook record available at https://lccn.loc.gov/2024014701

DESIGN AND PRODUCTION
Design by Rhea Magaro
Production by Beeline Media and Design, Inc.
Art direction by Tom Morgan

PHOTOGRAPHS by Alamy Stock Photos/Don Johnston, 7; Dreamstime/Chris Brignell, cover, Dssimages, 9, Geoffrey Kuchera, 10, Holly Kuchera, 8, Matthijs Kuijpers, 5; Shutterstock/Betty Shelton, 6, Bohbeh, 13, Eric Isselee, 11, Nynke van Holten, 2, 4, 12, 14

Printed in China